# Mountain Texture
## Glaciers of the Alps

Copyright © 2019 Garrett Fisher. All rights reserved.

ISBN: 9781675981535

Published by Tenmile Publishing LLC - USA
Website & Blog: garrettfisher.me

**Front Cover:** Kanderfirn, Switzerland.  **Above:** Bietschgletscher, Switzerland.  **Rear Cover:** Aletschgletscher, Switzerland.

▼ *Aletschgletscher, Switzerland.* ▲ *Dischliggletscher, Switzerland.* ▼ *Brunnenfirn, Switzerland.*

▼ *Triftgletscher, Switzerland.*

▼ Aletschgletscher, Switzerland.   ▲ Stuefesteigletscher, Switzerland.   ▲ Glacier de Valsorey, Switzerland.

▼ Aletschgletscher, Switzerland.   ▼ Plaine Morte Glacier, Switzerland.

▲ *Glacier de Corbassière, Switzerland.* ▼*Doldenhorngletscher, Switzerland.* ▶ *Glacier d'Otemma, Switzerland.*

# GLACIAL TEXTURE FROM ABOVE

"Texture" as I call it, captured from above in an antique airplane, has developed into a greater odyssey than I ever could have expected. In short, it was in 2012 that I happened upon surprising patterns tilled into Carolina red soil, while I was on the way to the Great Smoky Mountains. That fateful flight led me to discover that, if I used the zoom lens and focused on patterns, or what I later came to call "textures," I could find an element of beauty entirely different from my mainstay, which was scenic landscapes.

While still pursuing such things as national parks and massive mountains in the airplane, I kept an eye for agricultural patterns, collecting images as I traversed America. Eventually after relocating out West, I expanded my focus to include natural patterns and textures that had nothing to do with agriculture. Whether it was water, rocks, forests, canyons, snow, or other surfaces, there was an abundance of intriguing things that fit this pattern. Two books resulted from those efforts.

Some years later, the airplane made its way to Europe, where I continued the pursuit of texture. I have to admit I got a little scattered,

**ABOVE:**

*Gornergletscher above Zermatt, Switzerland.*

**RIGHT:**

*Glacial lake beneath Gruebenjoch, Switzerland.*

as European villages, farmland, and vineyards are mind blowing in their surreal beauty, with great variety depending on the country in question. I began collecting whatever I could find, expanding an ever greater list of subjects that I intended to write a book about.

What coalesced next was a texture book about the Pyrenees, of all things. As I was living in Spain in the mountains, I spent a lot of time wandering around that mountain range, branching into a new approach, which was to isolate texture to a specific geographic area. Prior to that project, I had collected images of a certain theme from an entire continent.

The problem was, I still struggled to explain what "texture" even was in this context. Three chapters of writing in the Pyrenees book, along with a large pile of compelling images still left me struggling to even identify what exactly the genre was supposed to be. Was it mathematical patterns? There was certainly a "requirement" to keep the horizon out of it– is that it, to take the horizon out and include a pattern of some sort?

I found the definition wanting, drifting into explanations about the

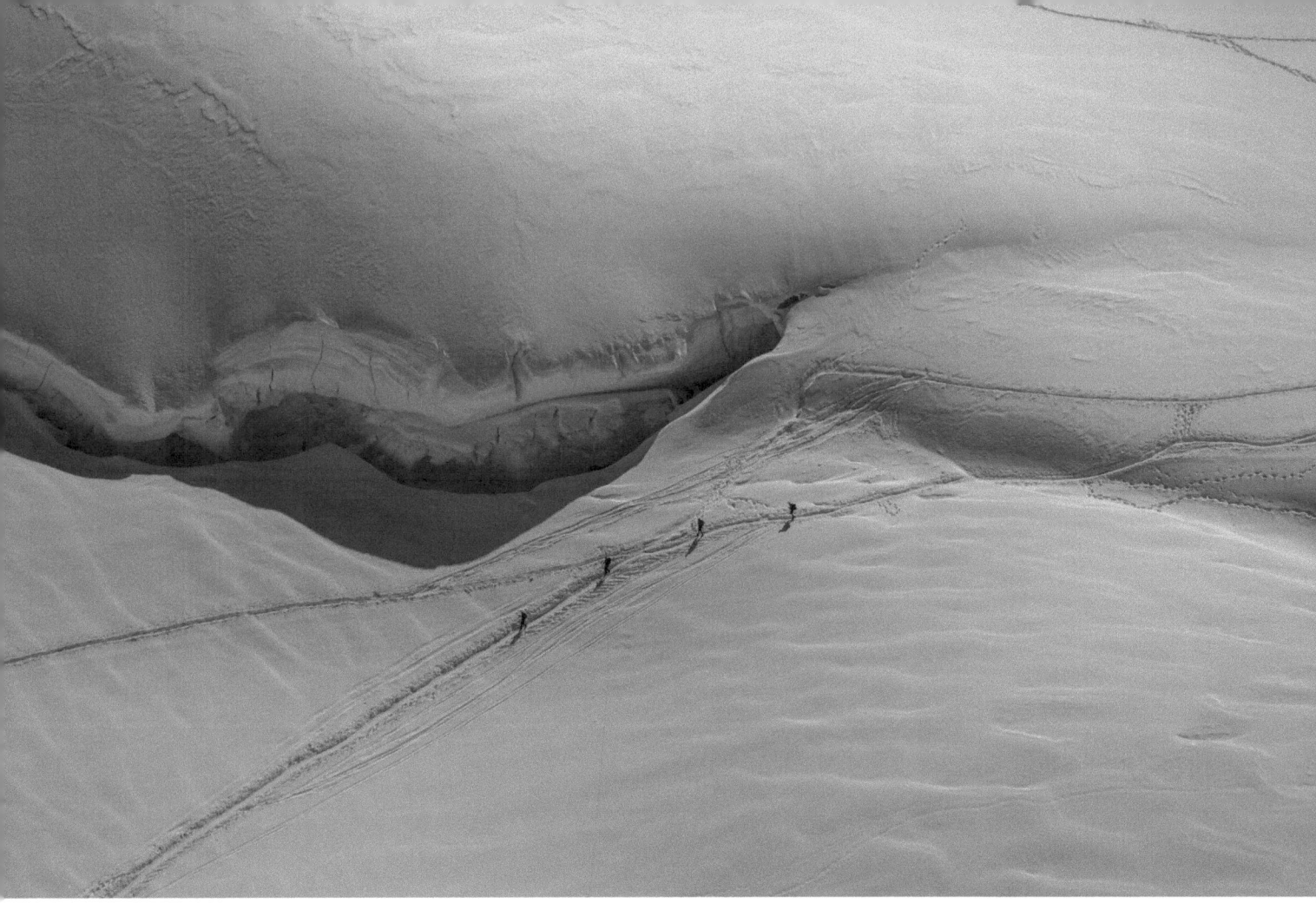

ABOVE:

*Ice cap on Parrotspitze, Switzerland.*

FAR MIDDLE LEFT:

*Grenzgletscher, Switzerland.*

FAR UPPER LEFT:

*Gornergletscher, Switzerland.*

LOWER LEFT:

*Aletschgletscher, Switzerland.*

UPPER CENTER:

*Seewjinengletscher, Switzerland.*

brilliance of the textured Spanish sky, illustrious colors, and other such realities that pervaded the area where I found my subjects. Was it that the place had special ingredients in its lighting, vegetation, and color or that I could find more patterns? Is not texture just the nature of the surface of something?

It was later that I would grab a magazine in Switzerland and note what a French photographer, who had photographed some scenery in the same fashion in Iceland, explained. By removing objects that could aid in identifying scale, the viewer would struggle to understand what the subject in the image actually is. The visual is set adrift from reality, which accomplishes an interesting outcome: the viewer is invited into a process to deduce what he or she is looking at, while at the same time enjoying an image that is aesthetically pleasing. Ironically, images of nearly epic scenic value are a form of spoon feeding of visual art, as the subject is instantly understood. Nonetheless, what I call "texture" is an image often with a pattern, bereft of horizon, and containing elements that make scale challenging to determine.

The Alps figure into this equation in another dimension that again operated outside of my expectations. My initial foray into the central Alps was for three months. While I had intended to eventually relocate there, I have a philosophy where I do not expect to be in the same place, with the same lighting and similar subject again. It is certainly possible to replicate elements of an image; however, the vast amount of variables necessary to coordinate in order to produce a flight at glacier levels, in an old airplane, in a massive mountain range, in a foreign country on another continent, with good weather and lighting is something that I do not take for granted. Thus, I decided to unleash and photograph everything I could find.

I expected to find some glacier texture images; I certainly expected to take many photographs of landscapes with glaciers in them. My primary focus was the peaks over 4000 meters, which often had glaciers as part of the scene, though I wasn't sure what I would do beyond the peaks project. Soon after wandering the Alps, I noticed that the place did not have the same amount of texture as the Pyrenees, requiring about six weeks of flying before I discovered that the most glaciated regions have a binary surface that often lacks texture that I was accustomed to seeing. Once I drifted out of the Valais in Switzerland, where glaciers and mountains are highest, I found slivers that reminded me of Spain, though I would categorize the textural result that "one could find some texture in the Alps if one looks hard enough."

The glaciers turned out to be a surprise. While I had expected some basic texture as seen from satellite images, it turned out to be a wonderland of crevasses, wrinkles, grain patterns, melt patterns, water movement, streams of rocks…I could go on. I believe that I failed to fully comprehend that a glacier is de-

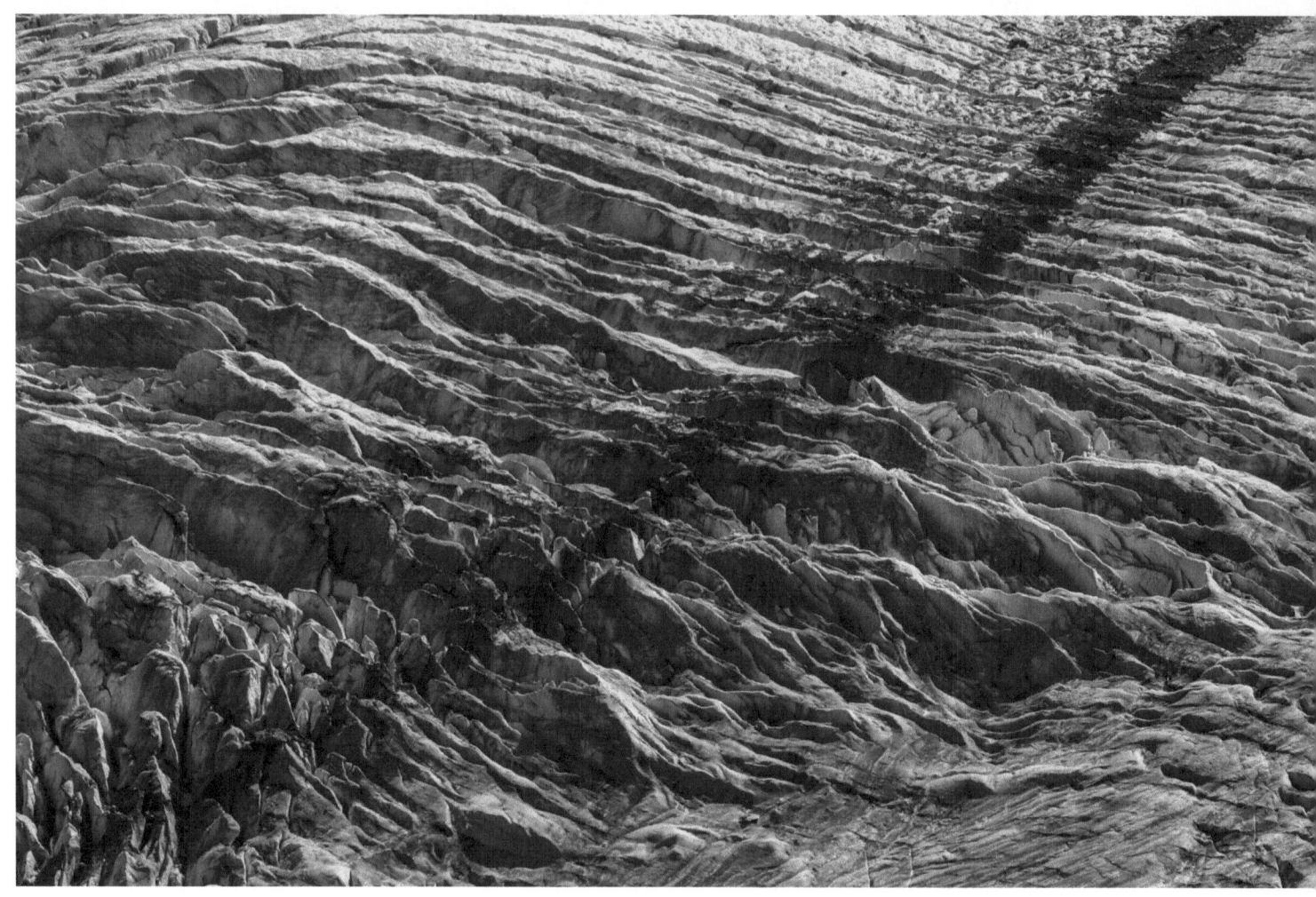

fined as a moving river of ice, as opposed to a "perennial ice field." Remaining glaciers in America and the Pyrenees are so small that, while they may move by definition, one does not sense it from the airplane. Instead, it looked at the time like "centuries of snowfall," which was emotionally stirring in its own right, to be staring at evidence of history through unmelted winters. The Alps feature glaciers of such sizes that it made my American and Iberian exploits look like snowdrifts.

Some glaciers move up to a half meter per day during the summer season. Yet, if one were to stand alongside it, it would be difficult to specifically note movement. Thus, a glacier is something of paradox, as it is a "river" of solid objects that moves while appearing not to move. It reminds me to some extent of the ocean, as when one stands on the coastline, it is in eternal motion, yet the location of the coast rarely moves at all. Nonetheless, there is a certain majesty standing on the ground facing a glacier, much like standing facing the ocean – it is a part of our natural ecosystem with tremendous power.

As I first uncovered the magnitude of glacial detail, I was a kid in a candy store, photographing as much as I could. Then I considered the bleak reality that the glaciers are melting at a tremendous rate of speed, in some instances the reality of said change taking place before my very eyes with thundering waterfalls bursting from beneath the glaciers, or recent icefalls cascading below the tongue. I contemplated the idea of documenting them before they disappear which, while it is something that motivates me and I do it somewhat for posterity, I can't help but reflect on the fact that scientists are doing a fantastic job of monitoring glaciers to a precision.

**ABOVE:**

*Unterer Theodulgletscher, Switzerland.*

**RIGHT:**

*Glacier du Mont Collon, Switzerland.*

It was not until I read an article in 2019 from one of Switzerland's prestigious universities that another perspective materialized. In 100 years, there may be only two glaciers left in Switzerland. Right now, there are hundreds, if not thousands! When playing around glaciers, climate change is a looming reality in the back of one's mind, though there was this mental concept that the worst of it would be so far off that I didn't have to think too hard about it. Now, in the course of a century, everything glacial that I am looking at could be gone. All of it.

I couldn't forget that bit of information, and that led to the creation of this book. I had many, many things to choose from to write, with well over 100,000 photos of other subjects long overdue to see the printed page. The reality struck me that, not only am I documenting the glaciers from my aerial perspective before they disappear, I am creating art that will not exist in a century.

Glacial texture is a natural form of art, turned into something consumable by humans through the technology, perspective, and medium made available since the late 19th century. Photography and aviation are relatively recent inventions, appearing when the Industrial Revolution began to get going in full steam. These tools make what I am doing possible; otherwise, it was the domain of birds to look into the heart of glacier as I can now. Yet, in a sad twist of irony, that same Industrial Revolution that brought these great tools is the same mechanism that is accelerating the demise of these wondrous natural features. I truly am racing to create art before it disappears for millennia.

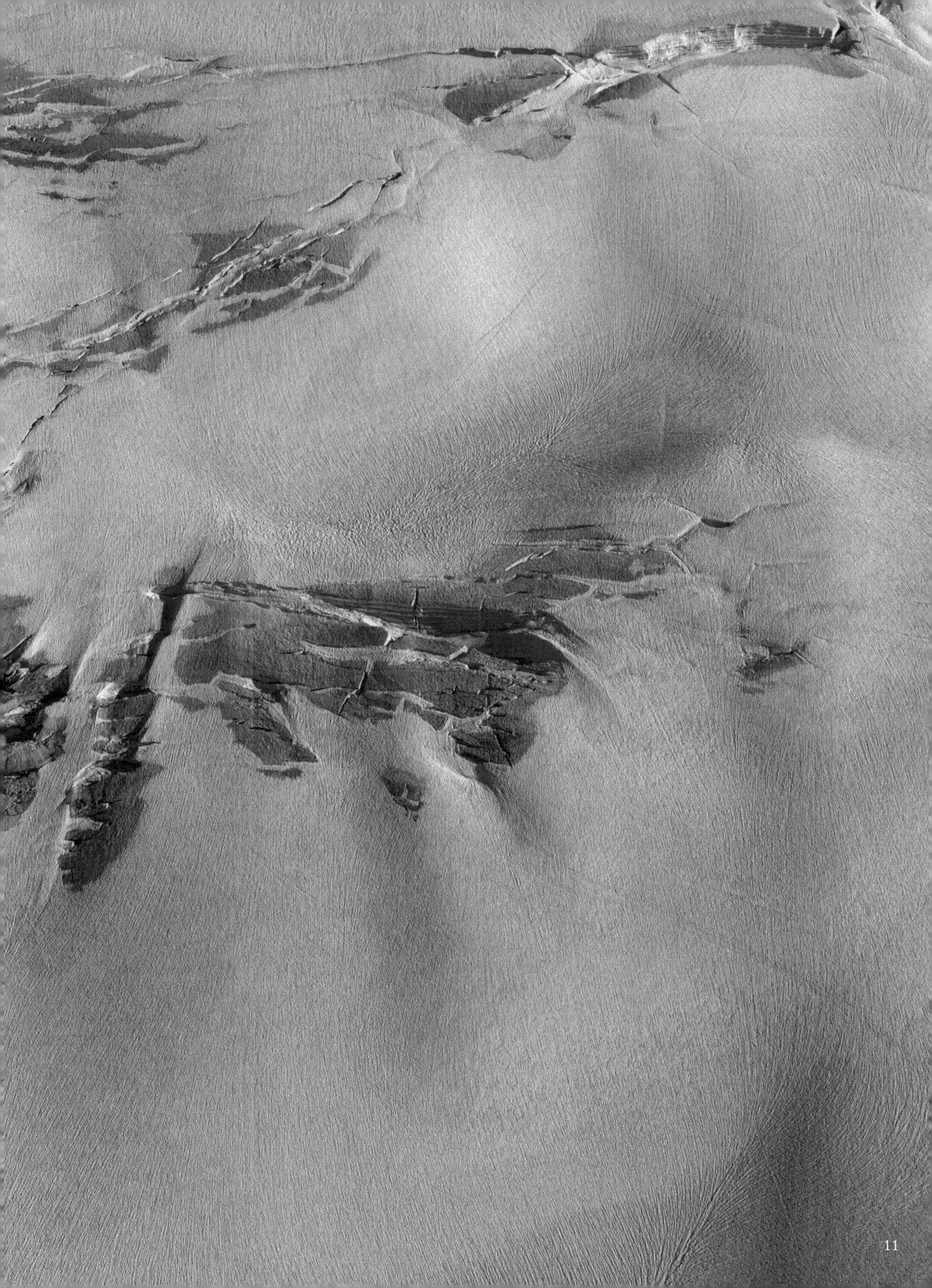

◄ *Feegletscher, Switzerland.* ▲ *Gamchigletscher, Switzerland.* ▼ *Vadrettin da Misaun, Switzerland.*

**ALL:** *Fiaschergletscher, Switzerland.*

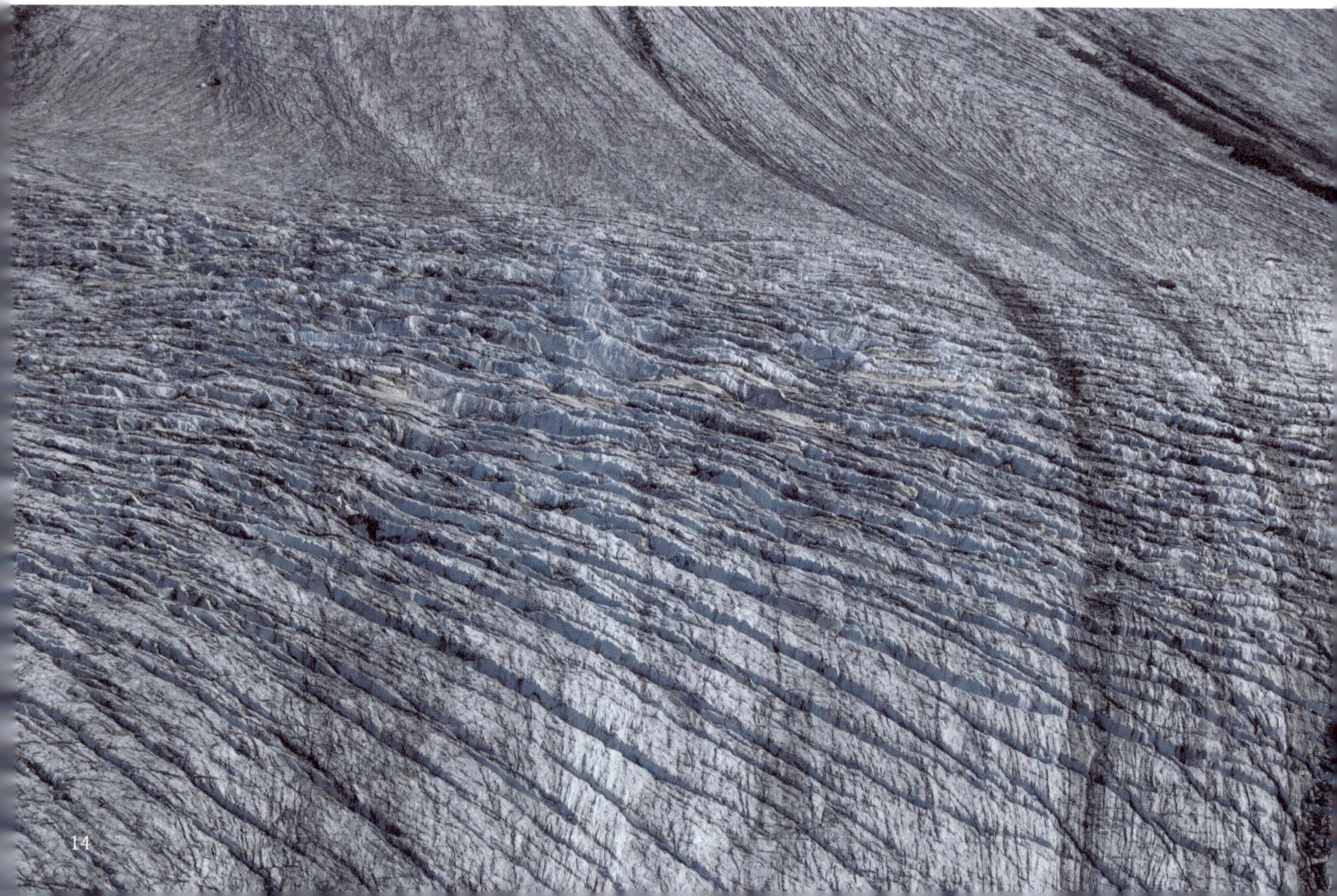

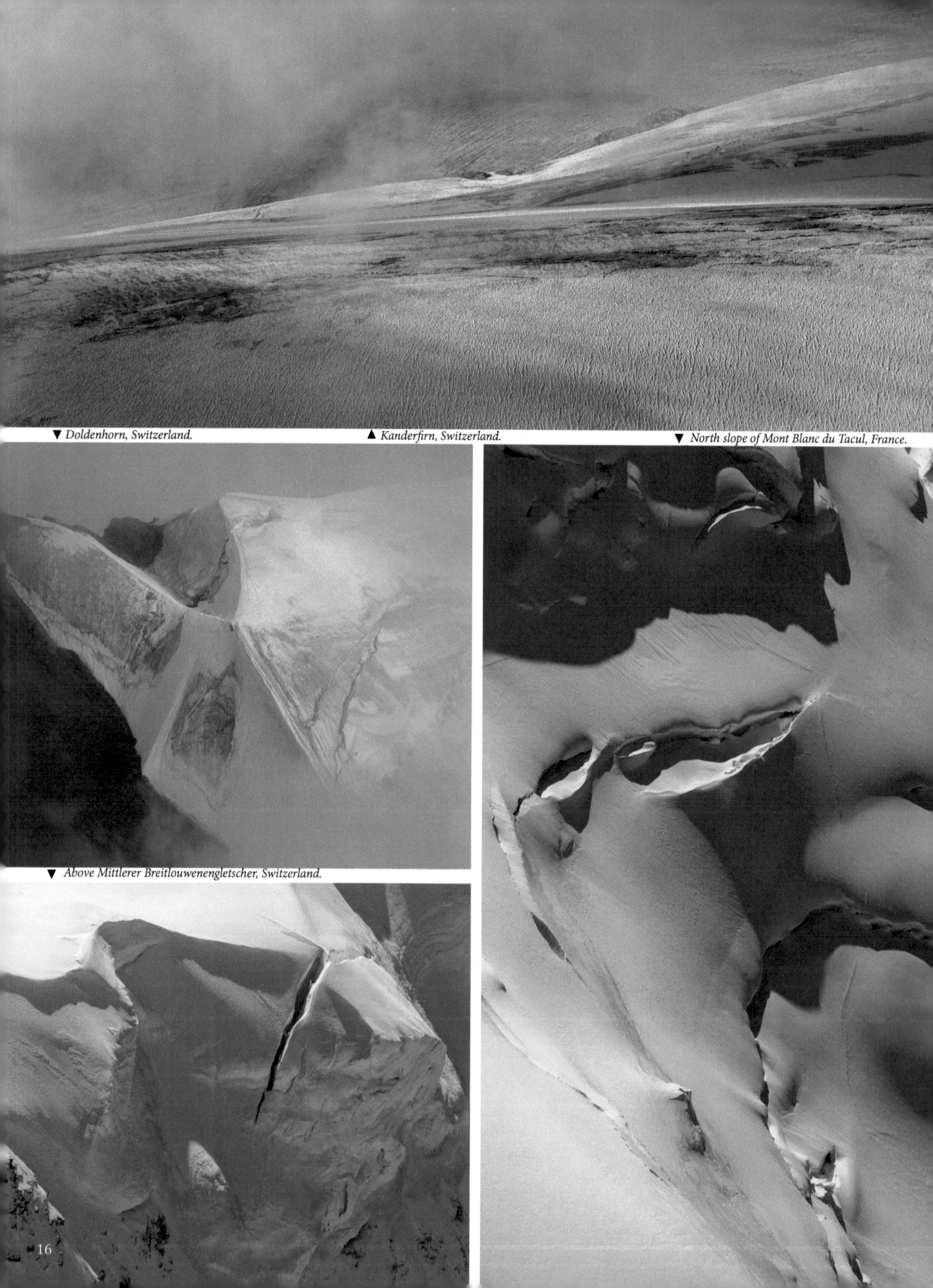

▼ *Doldenhorn, Switzerland.* ▲ *Kanderfirn, Switzerland.* ▼ *North slope of Mont Blanc du Tacul, France.*

▼ *Above Mittlerer Breitlouwenengletscher, Switzerland.*

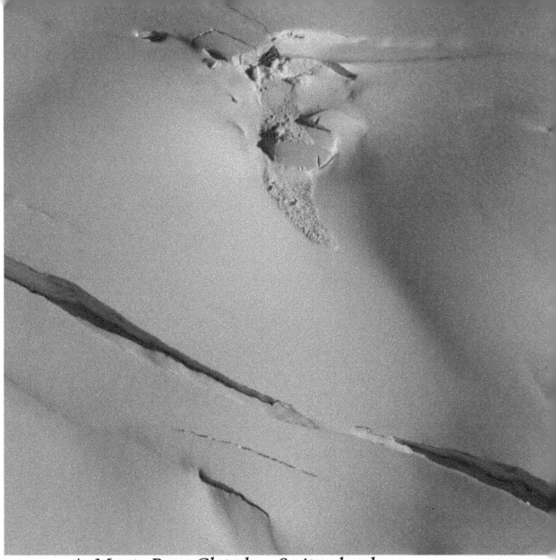

▼ *South slope of Doldenhorn, Switzerland.*  ▲ *Beneath Mont Maudit, France.*  ▲ *Monte Rosa Gletscher, Switzerland.*

▼ *Glacier de Taconnaz, France.*  ▼ *Zwillingsgletscher, Switzerland.*

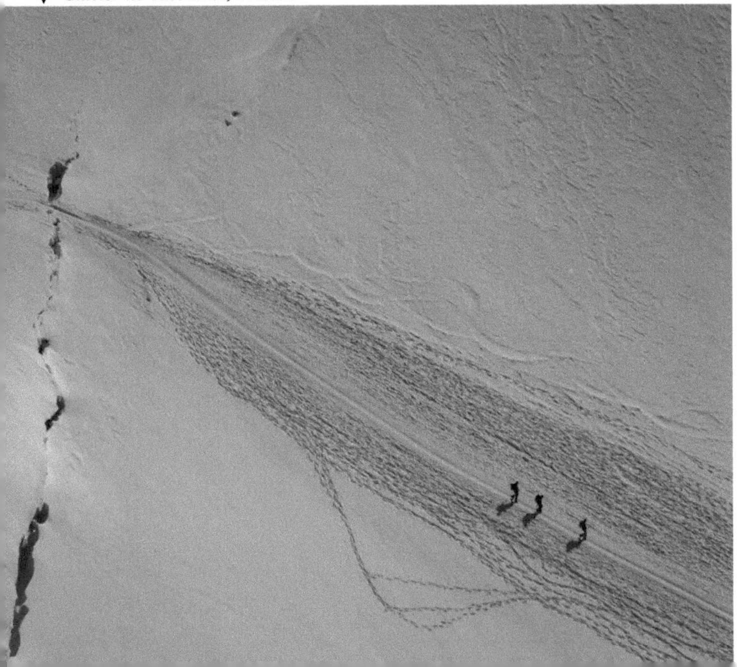

◀ *Glacier du Grand Combin, Switzerland.* ▲*Wyssenbachgletscher, Switzerland.* ▼*Beneath Fletschhorn, Switzerland.*

▲ Üsser Talgletscher, Switzerland.    ▼Tierberggletscher, Switzerland.    ▶Plaine Morte Glacier, Switzerland.

◄ Aargletscher, Switzerland. ▲Finsteraargletscher, Switzerland. ▼Oberaargletscher, Switzerland.

▲ Konkordiaplatz - Aletschgletscher, Switzerland.   ▼Rhône Glacier, Switzerland.   ► Aletschgletscher, Switzerland.

▼ *Aletschgletscher, Switzerland.*  ▲ *Fieschergletscher, Switzerland.*  ▼ *Aletschgletscher, Switzerland.*

▼ *Allalingletscher, Switzerland.*

▼ *Fiechergletscher, Switzerland.* ▲ *Aletschgletscher, Switzerland.* ▲ *Aletschgletscher, Switzerland.*

▼ *Aletschgletscher, Switzerland.* ▼ *Glacier d'Argentière, France.*

◀ *Glacier du Trient, Switzerland.* ▲ *Glacier d'Argentière, France.* ▼*Hohwänggletscher, Switzerland.*

▲ Riedgletscher, Switzerland.   ▼Schwärzegletscher, Switzerland.   ►Riedgletscher, Switzerland.

◀ *Gauligletscher, Switzerland.* ▲ *Ischmeer, Switzerland.* ▼*Ischmeer, Switzerland.*

▲ Gornergletscher, Switzerland.   ▼ Plaine Morte Glacier, Switzerland.   ▶ Mittelaletschgletscher, Switzerland.

▼ Rutor Glacier, Italy.  ▲ Obers Ischmeer, Switzerland.  ▼ Innre Baltschiedergletscher, Switzerland.

▼ Fieschergletscher, Switzerland.

▼ *Planpincieux Glacier, Italy.* ▲ *Grenzgletscher, Switzerland.* 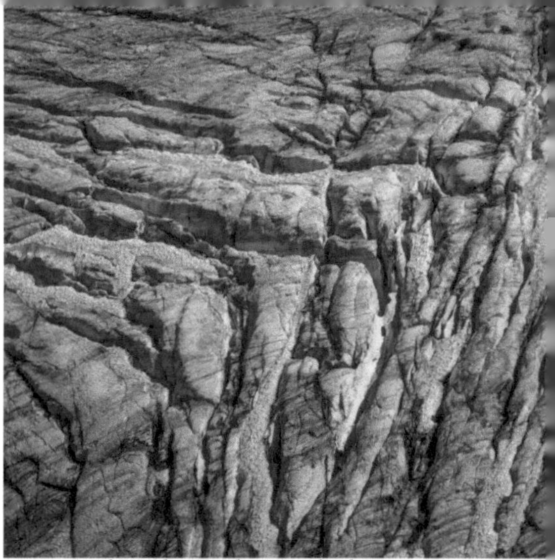 ▲ *Glacier des Grands, Switzerland.*

▼ *Glacier des Diablerets, Switzerland.*  ▼ *Vadret dal Chaputschin, Switzerland.*

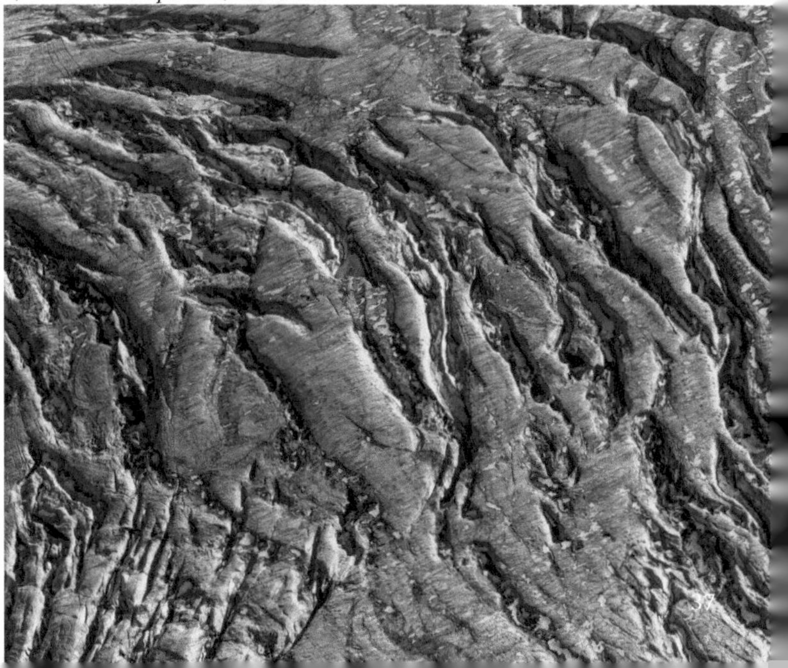

◀ *North of Bruneggletscher, Switzerland.* ▲*Ghiacciaio del Caoagnöö, Switzerland.* ▼*Aletschgletscher, Switzerland.*

▲ *Gornergletscher, Switzerland.* ▼ *Chilchligletscher, Switzerland.* ▶ *Gornergletscher, Switzerland.*

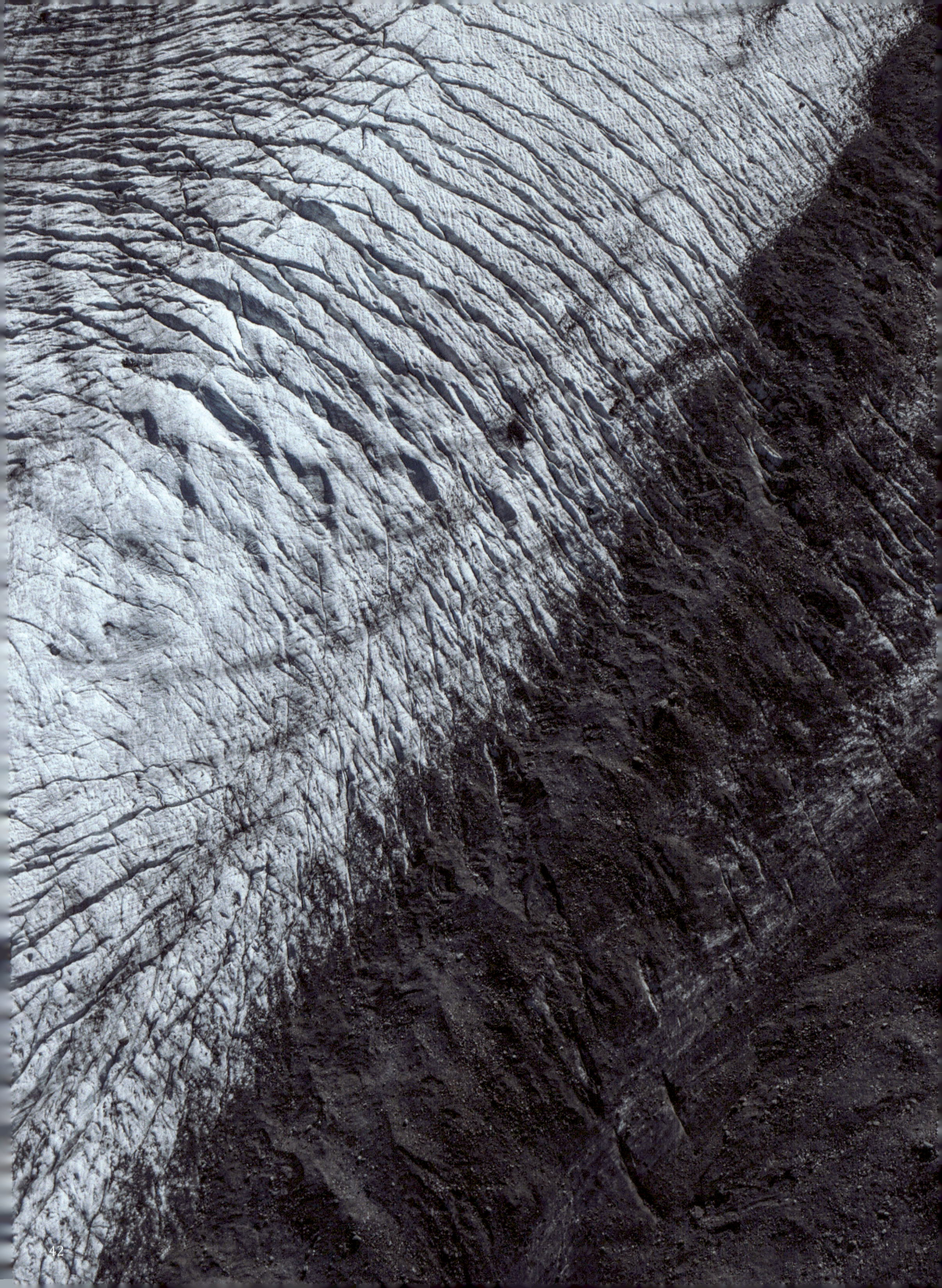

◄ *Morteratschgletscher, Switzerland.* ▲*Fluegletscher, Switzerland.* ▼*Doldenhorngletscher, Switzerland.*

▲ *Mittlerer Breitlouwenengletscher, Switzerland.*   ▼ *Slopes leading to Ischmeer, Switzerland.*   ▶   *Grosser Aletschfirn, Switzerland.*

▼ Glacier des Bossons, France. ▲ Blüemlisalpgletscher, Switzerland. ▼ Planpincieux Glacier, Italy.

▼ Eigergletscher, Switzerland.

▼ Bisgletscher, Switzerland. ▲ Hohwänggletscher, Switzerland. ▲ Glacier des Rognons, France.

▼ Glacier du Tour, France. ▼ Oberaletschfirn, Switzerland.

◂ *Feegletscher, Switzerland.*   ▴*Kranzbergfirn, Switzerland.*   ▾*Allalingletscher, Switzerland.*

▲ Glacier d'Argentière, France. ▼ Aletschgletscher, Switzerland. ▶ Aletschgletscher, Switzerland.

◄ *Birchgletscher, Switzerland.* ▲*Wildstrubelgletscher, Switzerland.* ▼*Triftgletscher, Switzerland.*

▲ Wetterlückengletscher, Switzerland.  ▼ Kranzbergfirn, Switzerland.  ► Kanderfirn, Switzerland.

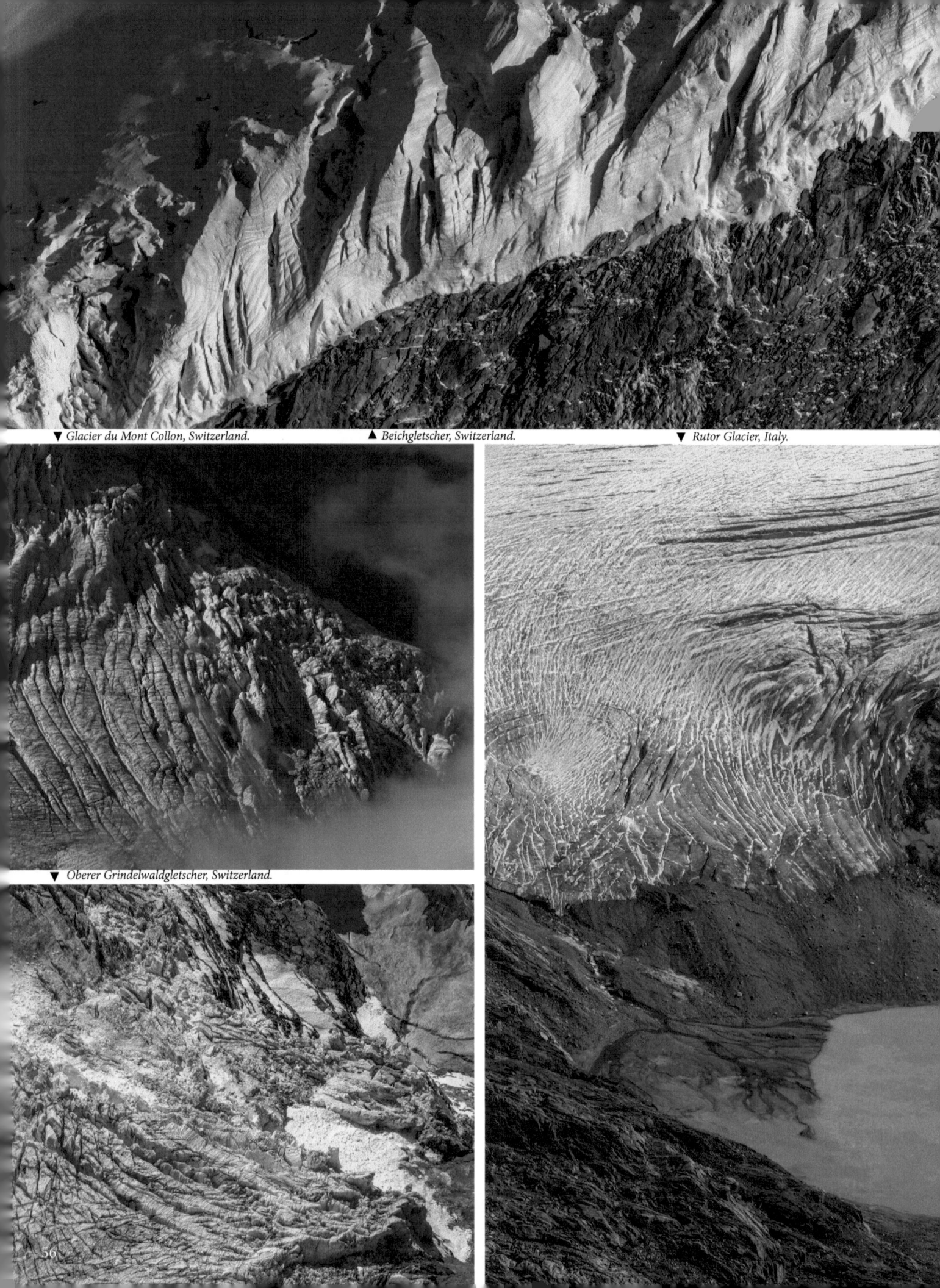

▼ *Glacier du Mont Collon, Switzerland.* ▲ *Beichgletscher, Switzerland.* ▼ *Rutor Glacier, Italy.*

▼ *Oberer Grindelwaldgletscher, Switzerland.*

▼ *Glacier du Mont Ruan, Switzerland.* ▲ *Glacier de Moming, Switzerland.* ▲ *Glacier du Trient, Switzerland.*

▼ *Beichgletscher, Switzerland.* ▼ *Glacier d'Argentière, France.*

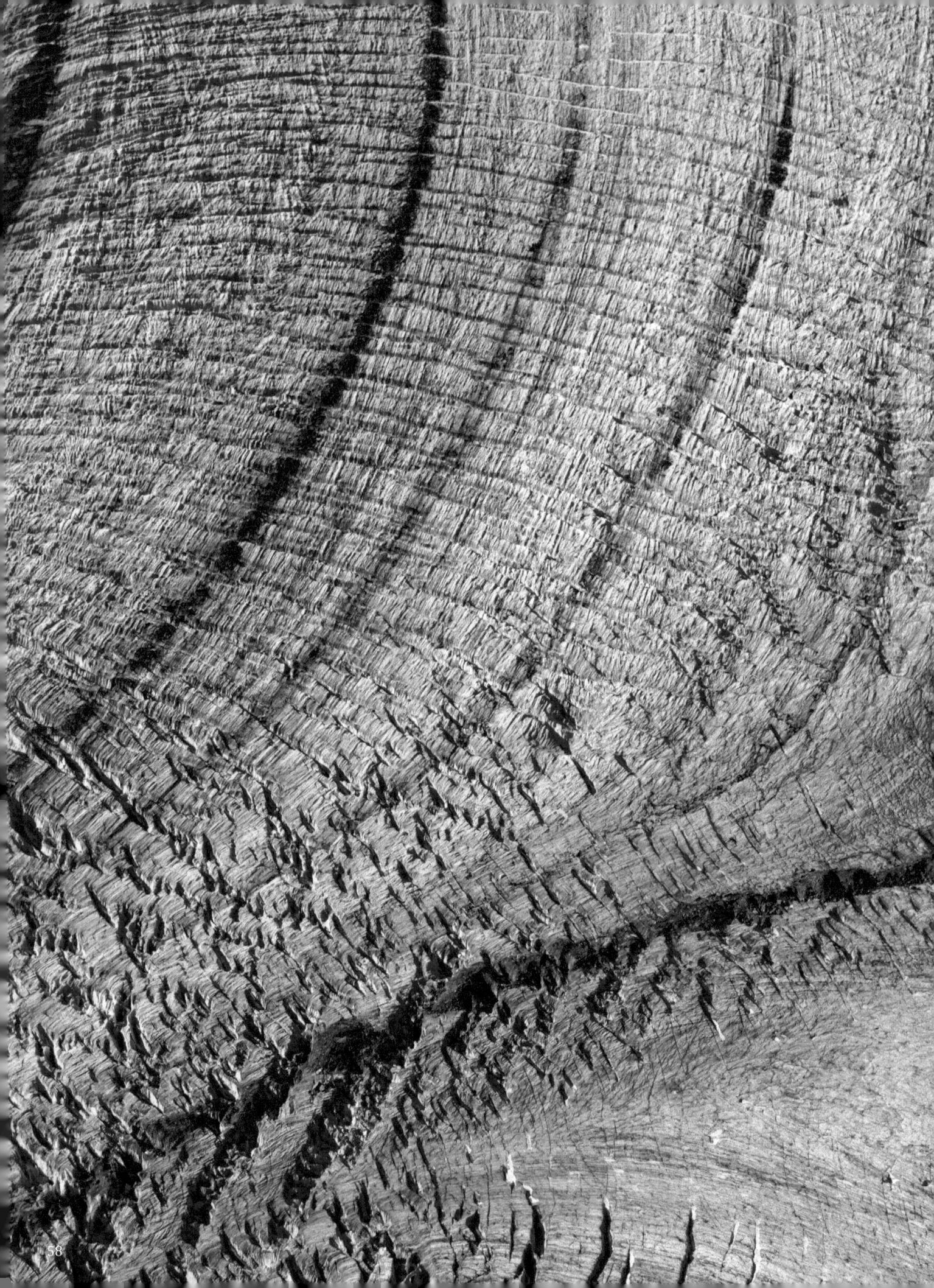

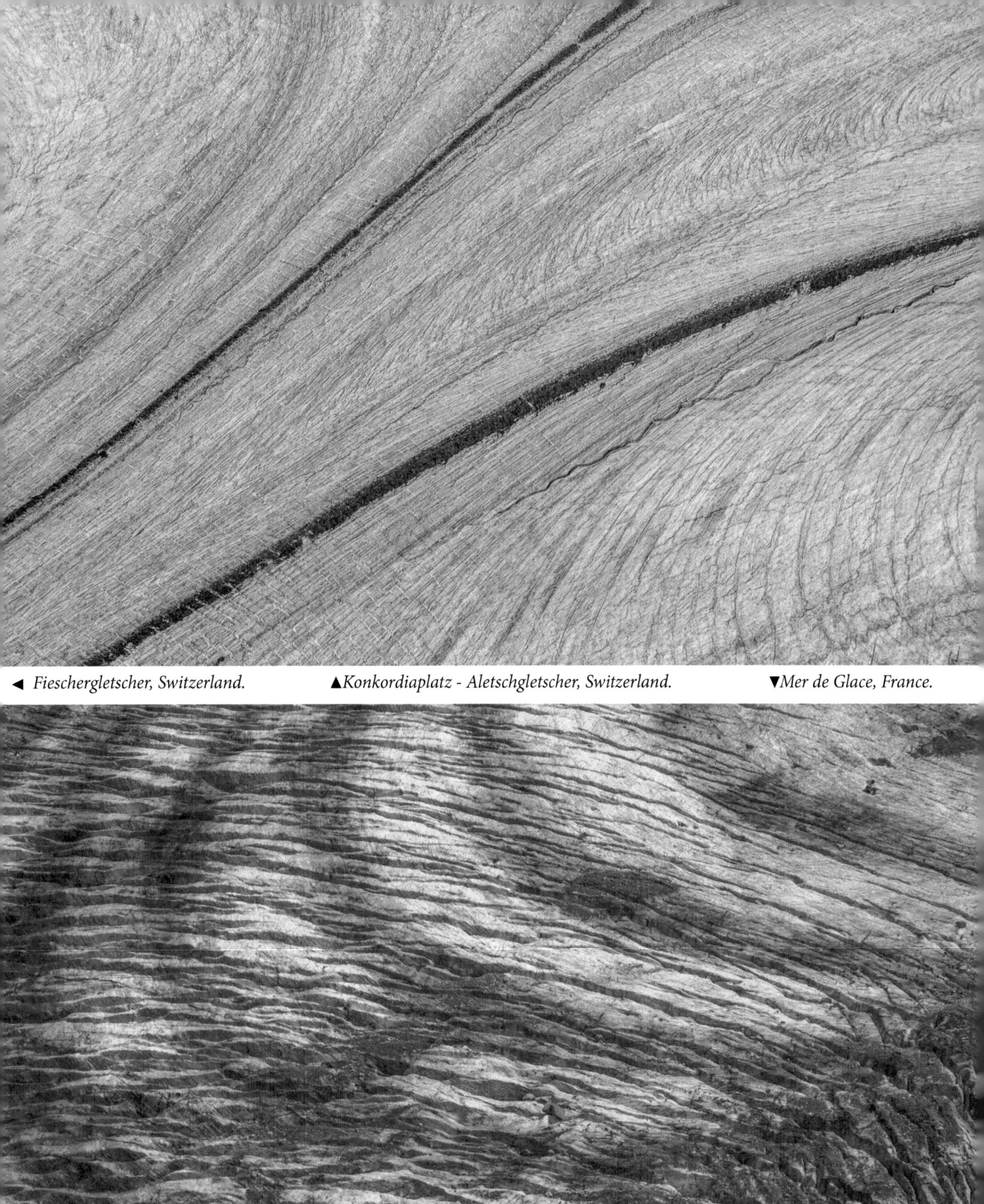

◀ *Fieschergletscher, Switzerland.* ▲*Konkordiaplatz - Aletschgletscher, Switzerland.* ▼*Mer de Glace, France.*

▲ *Glacier des Bossons, France.* ▼*Glacier de Bionnassay, France.* ► *Bisgletscher, Switzerland.*

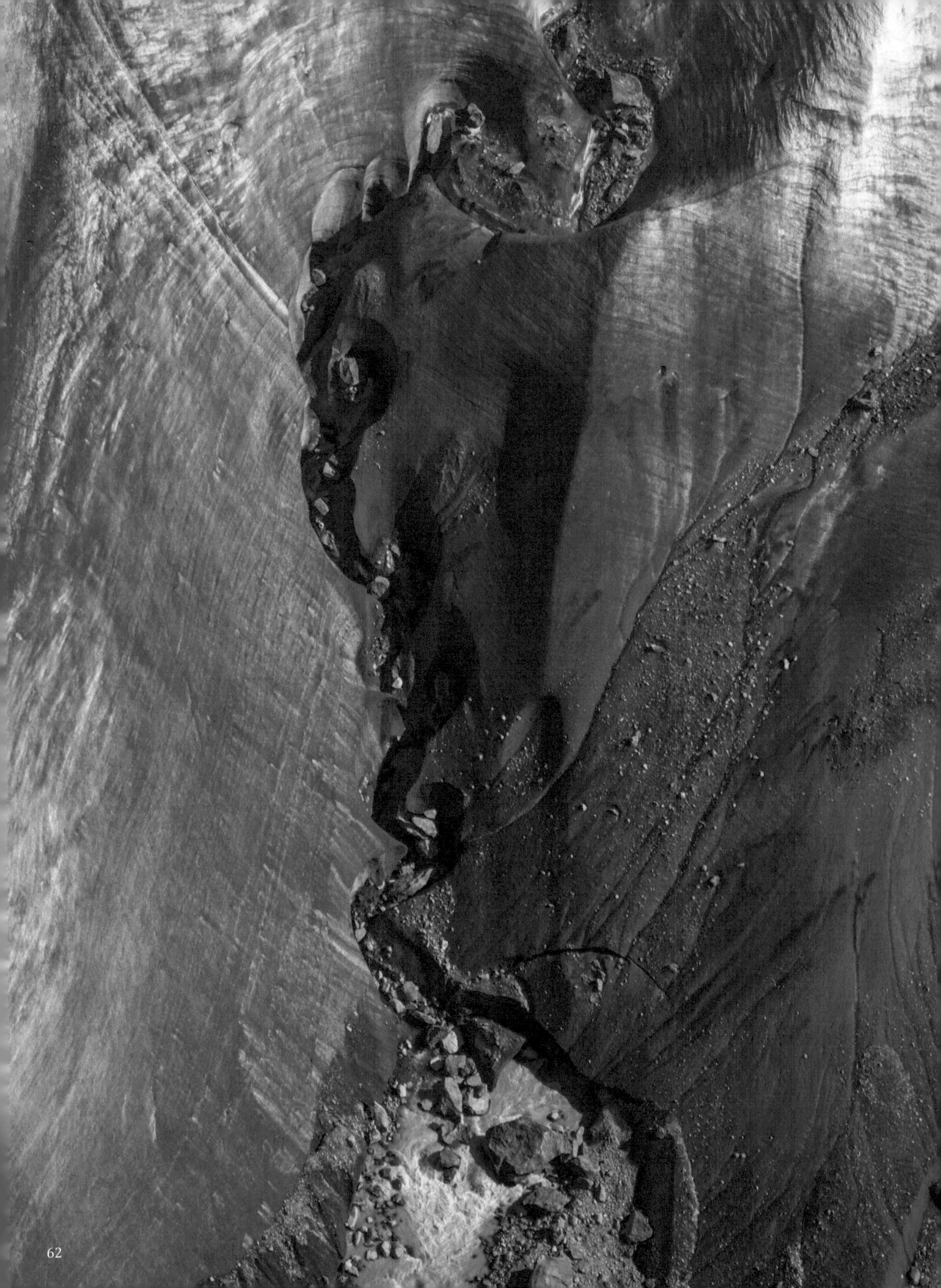

◄ *Tongue of the Gornergletscher, Switzerland.* ▲*Aletschgletscher, Switzerland.* ▼*Alpetilgletscher, Switzerland.*

▲ *Nördlicher Breitlouwenengletscher, Switzerland.* ▼*Jungfraufirn, Switzerland.* ▶ *Beichgletscher, Switzerland.*

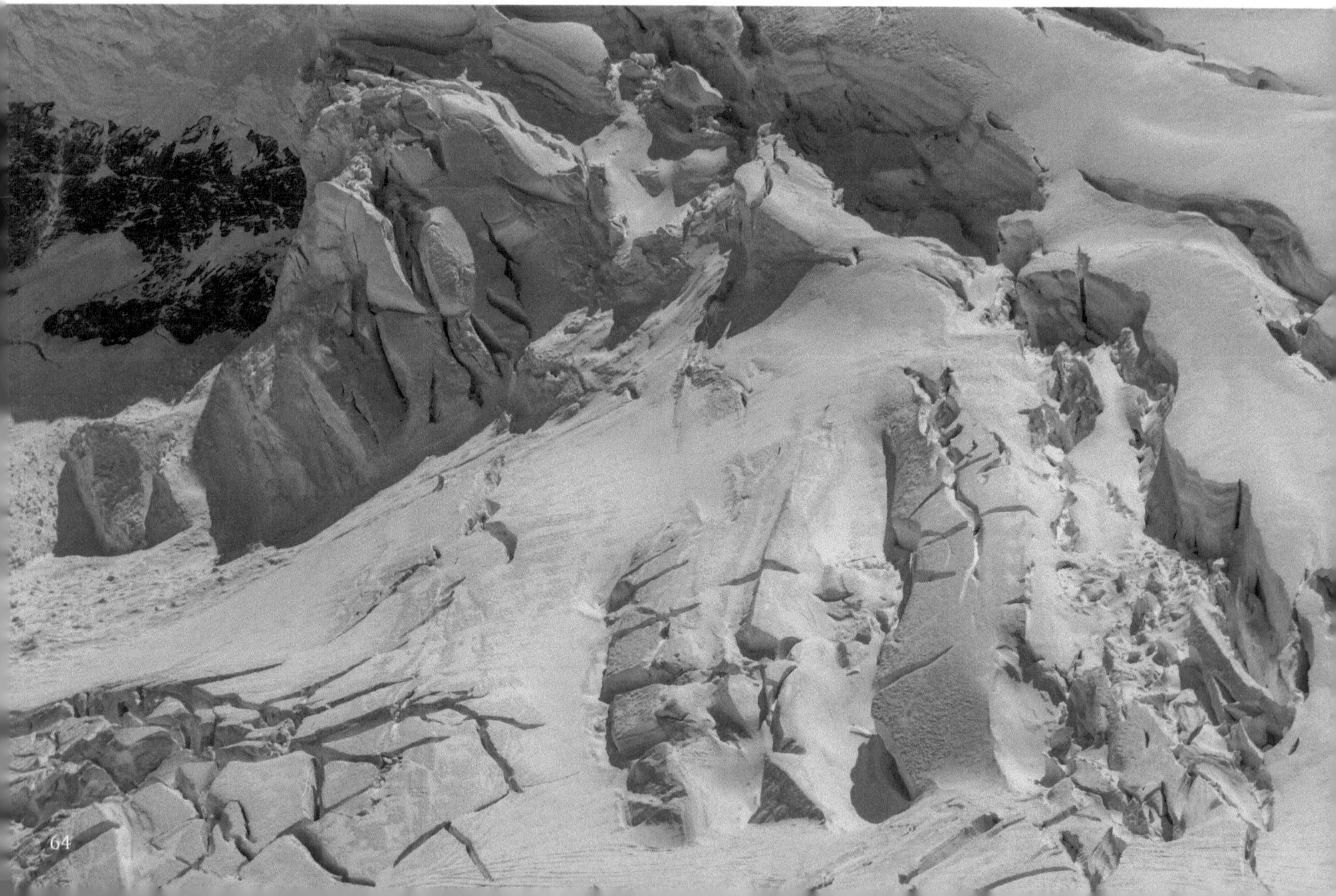

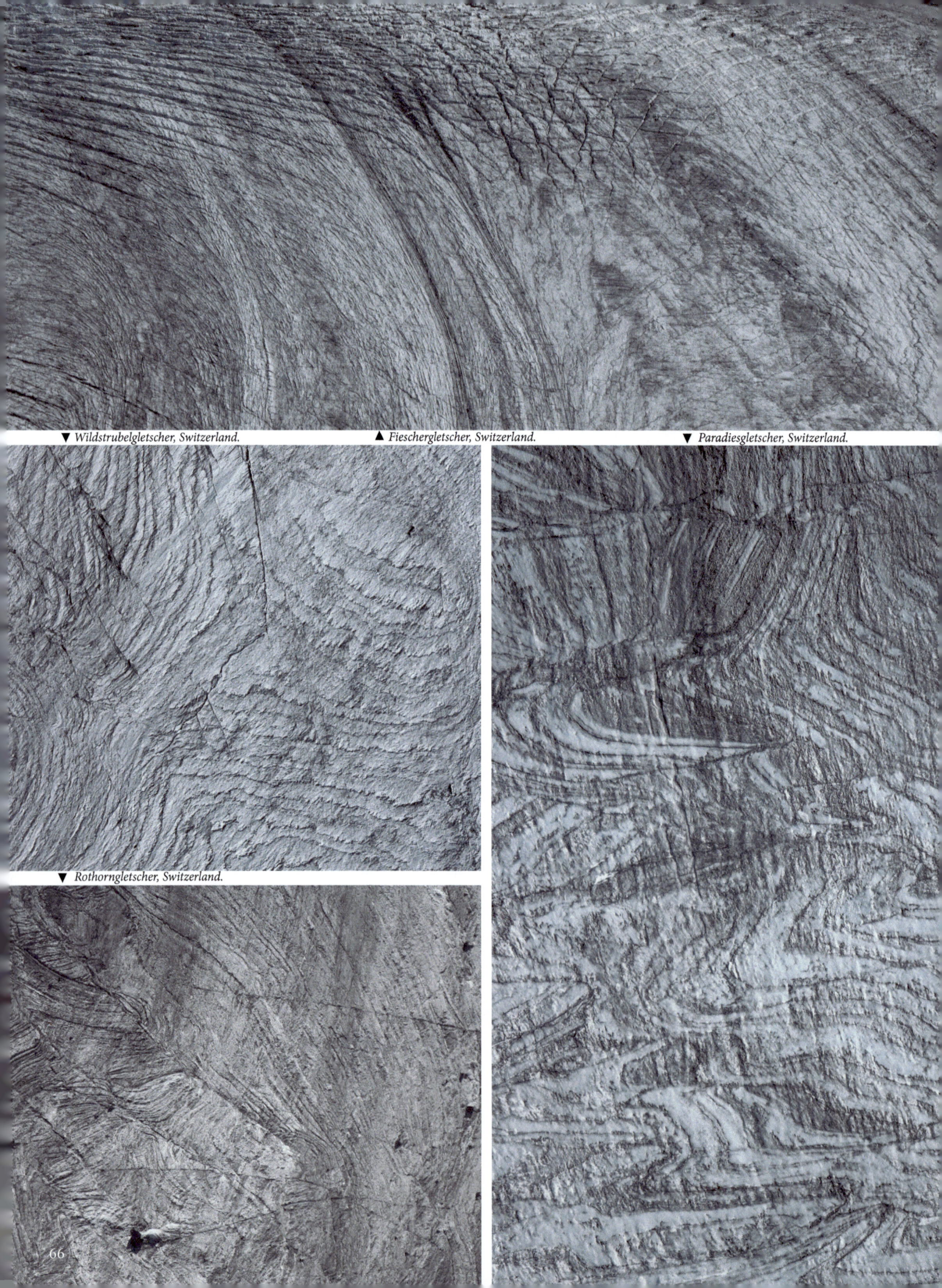

▼ Wildstrubelgletscher, Switzerland. ▲ Fiaschergletscher, Switzerland. ▼ Paradiesgletscher, Switzerland.

▼ Rothorngletscher, Switzerland.

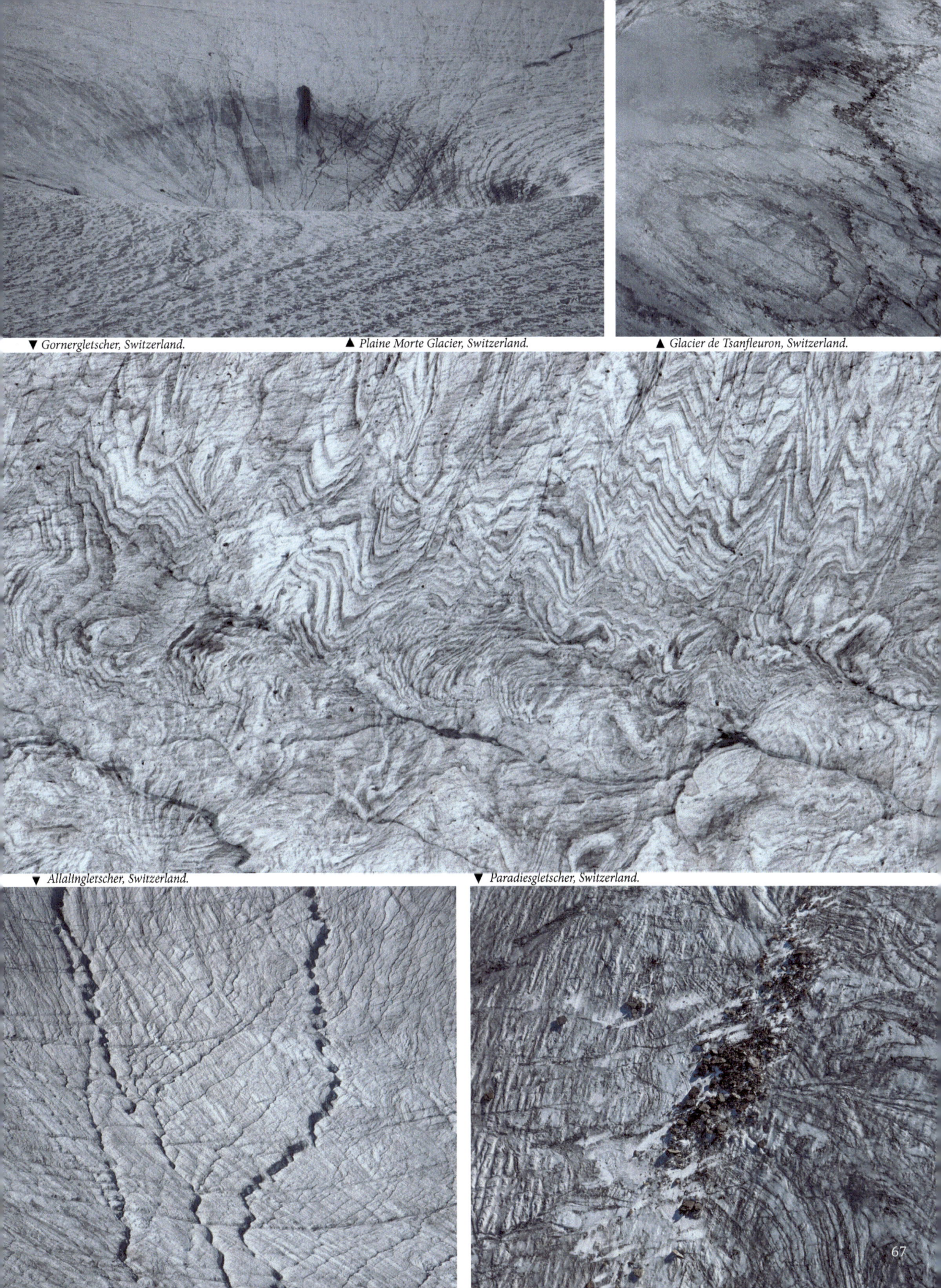

▼ Gornergletscher, Switzerland.　　▲ Plaine Morte Glacier, Switzerland.　　▲ Glacier de Tsanfleuron, Switzerland.

▼ Allalingletscher, Switzerland.　　▼ Paradiesgletscher, Switzerland.

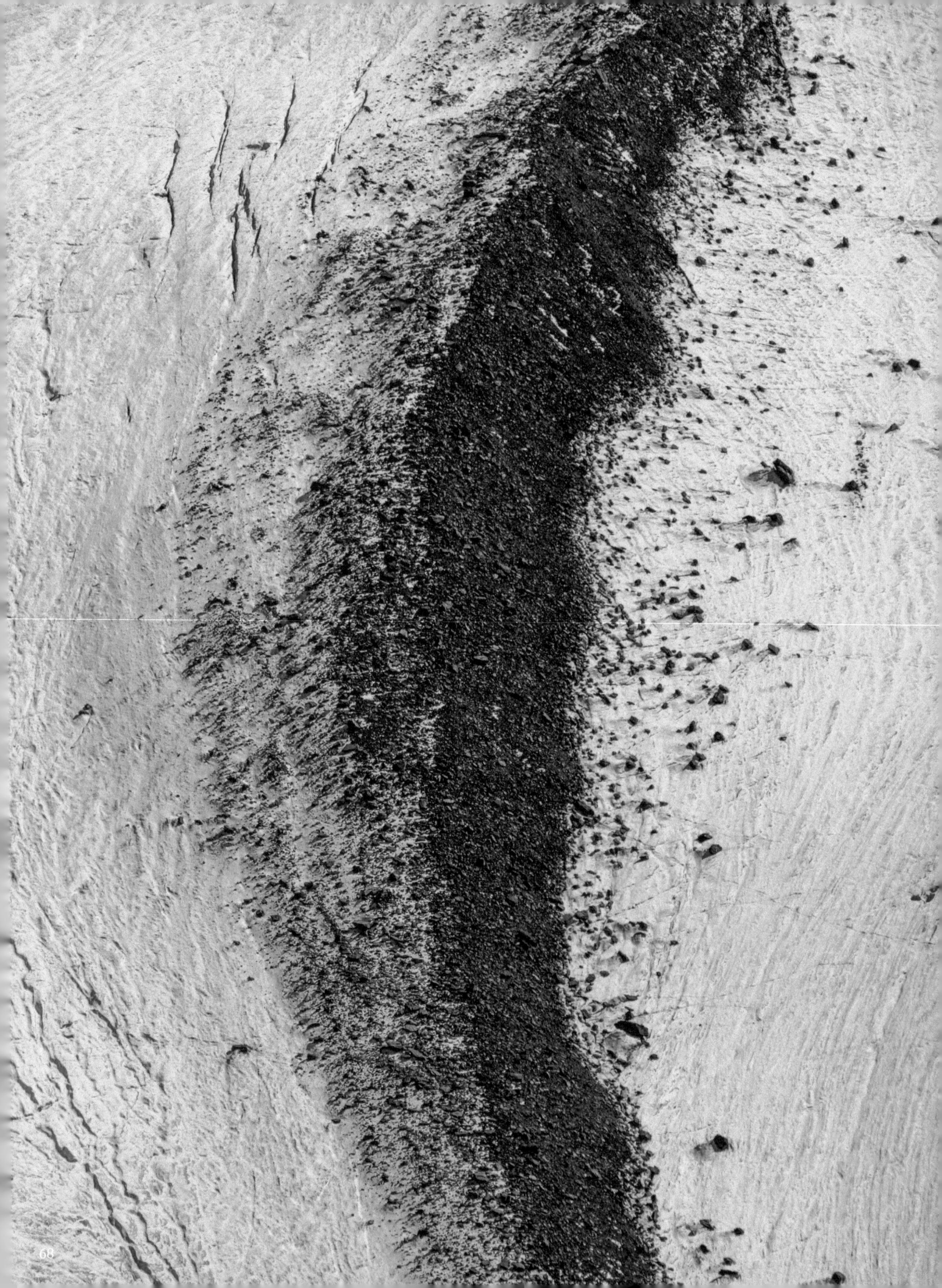

◀ *Rutor Glacier, Italy.*   ▲*Plaine Morte Glacier, Switzerland.*   ▼*Oberaargletscher, Switzerland.*

▲ Mont Maudit, France.  ▼ Jungfraufirn, Switzerland.  ▶ Grosser Aletschfirn, Switzerland.

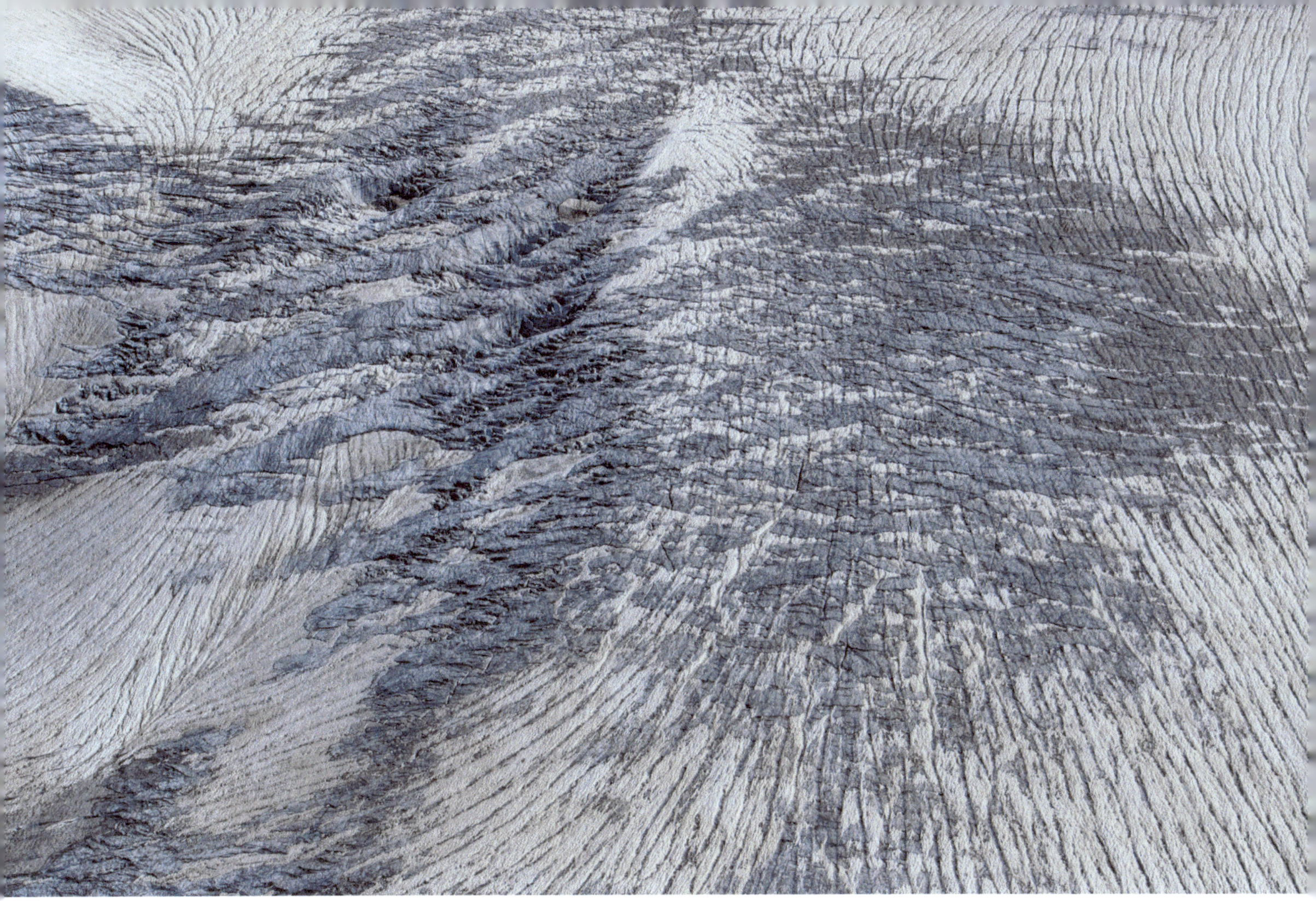

◄ *Mer de Glace, France.* ▲*Triftgletscher, Switzerland.* ▼*Glacier du Tour, France.*

**ALL:** *Glacier de Valsorey, Switzerland.*

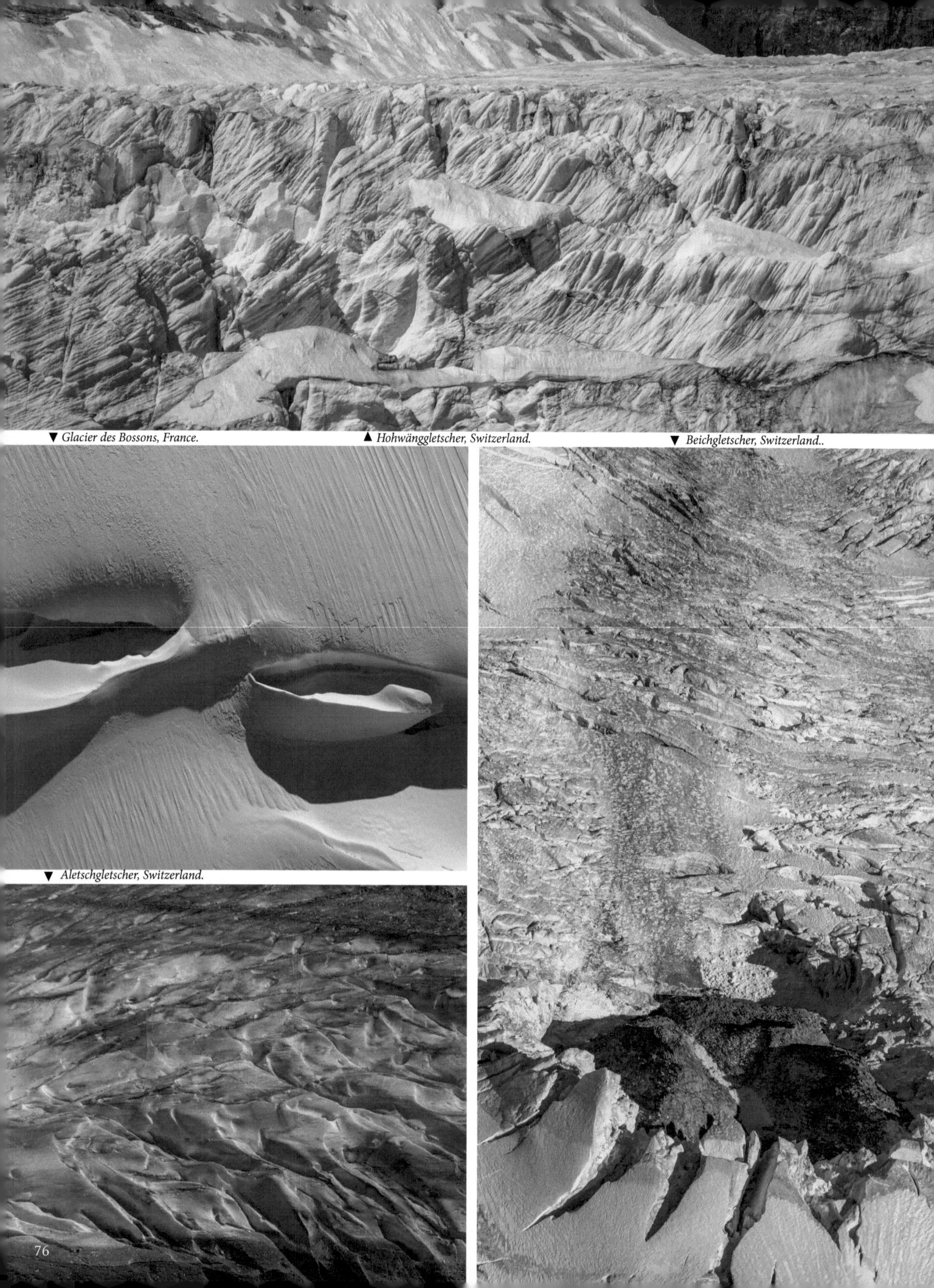

▼ Glacier des Bossons, France. ▲ Hohwänggletscher, Switzerland. ▼ Beichgletscher, Switzerland.

▼ Aletschgletscher, Switzerland.

▼ *Aletschgletscher, Switzerland.* ▲ *North side of Gspaltenhorn, Switzerland.* ▲ *Schwarzberggletscher, Switzerland.*

▼ *Konkordiaplatz - Aletschgletscher, Switzerland.* ▼ *Glacier d'Argentière, France.*

# Epilogue of the Glaciers

Are so many glacial images overwhelming? Imagine taking hours upon hours of flights over these majestic features, surrounded by technically challenging and inhospitable terrain, both for hikers and aviators. What is contained in this book is a minute fraction of the images taken mostly in 2018, with some in 2019. After applying what I thought was the most sensible curation method, I still uncover images that I believe belong in a collection of my best glacial textures. Yet, these features that appear in this book do not exist anymore, at least not in the exact form as presented in the book.

Depending on the speed of the glacier, it would have moved hundreds of feet since the time the image was taken, with patterns moving downhill, modifying as they interact with the mountainside and invisible rocks beneath the glacier. New rock falls and avalanches tumble onto the top of the glacier, depositing snow, soil, and rock to be carried downstream. Meanwhile winter adds its contribution above the firn line, leaving new snows to continue what remains of the cycle until they are gone.

While the overarching story of the glaciers' demise is in the forefront of my mind as I continue to explore and document them, I can't help but reflect on the fact that, during their recession phase, they are still producing incredible natural art and textures. Given the difficulty of traversing a glacier on foot due to the danger of deep crevasses, it appears that the approach from above, intimately overflying them, is something that is missed for favor of satellites and the occasional foray with a drone. I can truly say that, while the imagery produced is a treasure, there is little that can compare to what it's like being in the cockpit. When flying up a long, ascending valley with towering terrain on three sides, a large glacier beneath, smaller glaciers spilling down the side, and an ice-covered headwall in front of me, it feels like I am in forbidden territory, a place undiscovered that somehow nobody knows about. Perhaps it has something to do with the fact that, should the engine quit in this moment, the outcome would be egregious.

Nonetheless, the point remains that the existence of these natural features and the art that they produce will one day be looked upon by future generations of humans (if they still exist) as some form of prehistoric cave painting that came from a time so incredibly alien that it is hard to fathom. That is unless we do something about it and limit the damage, though I am not hopeful. Until then, I will keep documenting them.

**ABOVE:**

*Obers Ischmeer, Switzerland.*

**LEFT:**

*Plaine Morte Glacier, Switzerland.*

# More Books by the Author
## www.garrettfisher.me

Using the most ill-equipped aircraft possible for such adventure flying, Garrett Fisher has based his antique Piper PA-11 Cub Special in the Outer Banks of North Carolina, the highest airport in North America in Colorado, an elitist airpark near Yellowstone, the busiest airport in Germany, along the Coast of Portugal, at a secluded hideout in the Spanish Pyrenees, and at an astonishingly expensive airport in the Swiss Alps. In the process, he has flown to an exhaustive list of rugged and dangerous places in America as well as some of the most beautiful sites in Europe, amassing an enormous collection of aerial photographs. Perpetually clueless about what is coming next, he continues wandering in the airplane, blogging about his adventures at www.garrettfisher.me.

*Photo: Adam Romer*